World of science

ATMOSPHERE, SPACE AND SATELLITES

BAY BOOKS LONDON & SYDNEY

1979 Published by Bay Books
157–167 Bayswater Road, Rushcutters
Bay NSW 2011 Australia

National Library of Australia
Card Number and ISBN 0 85835 267 2
Design: Sackville Design Group
Printed by Tien Wah Press, Singapore.

THE EARTH'S ATMOSPHERE

The layer of air surrounding the earth is called the *atmosphere.* Although you cannot see, taste or smell air, it makes all life on earth possible. Without air to breathe, no plant, animal or human could live.

The air is kept from escaping out into space by the earth's *gravity,* the force which pulls everything down towards the planet. This gravitational pull gives everything on earth, including air, its weight.

Air presses on us all uniformly. Using an imaginary column of air, it has been calculated that air exerts a pressure of nearly 1kg per cm^2, at the earth's surface.

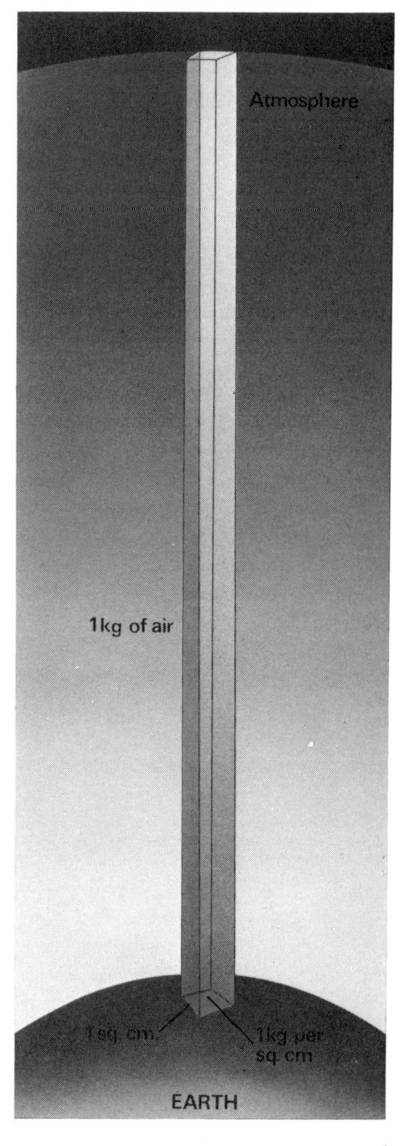

The air we breathe

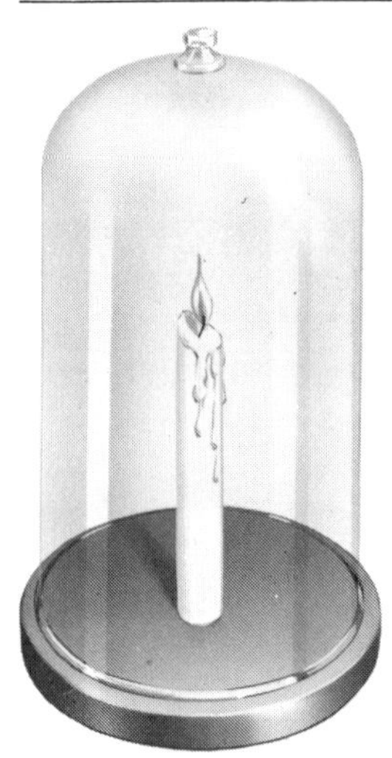

Without air there is no life. A lighted candle inside a bell jar will go out if all the air is pumped out. It cannot burn without oxygen.

Air consists of a mixture of gases. The most important of these are nitrogen and oxygen. Nitrogen makes up 78.09 per cent of air and oxygen 20.95 per cent. A gas called argon makes up 0.93 per cent and there are also very tiny amounts of carbon dioxide, helium, hydrogen, krypton, methane, neon, ozone and xenon, but they add up to only 0.03 per cent of air.

The composition of air can vary depending on where you are; this is especially true of the amount of carbon dioxide. When we breathe out, we *exhale* carbon dioxide so there is more in the air of crowded cities or places like closed rooms where the carbon dioxide cannot get out or new air in. This is why we call them 'stuffy' and look for windows to open to let in fresh air.

Although carbon dioxide is no use to human beings it is an important gas. Green plants use it to live in a process

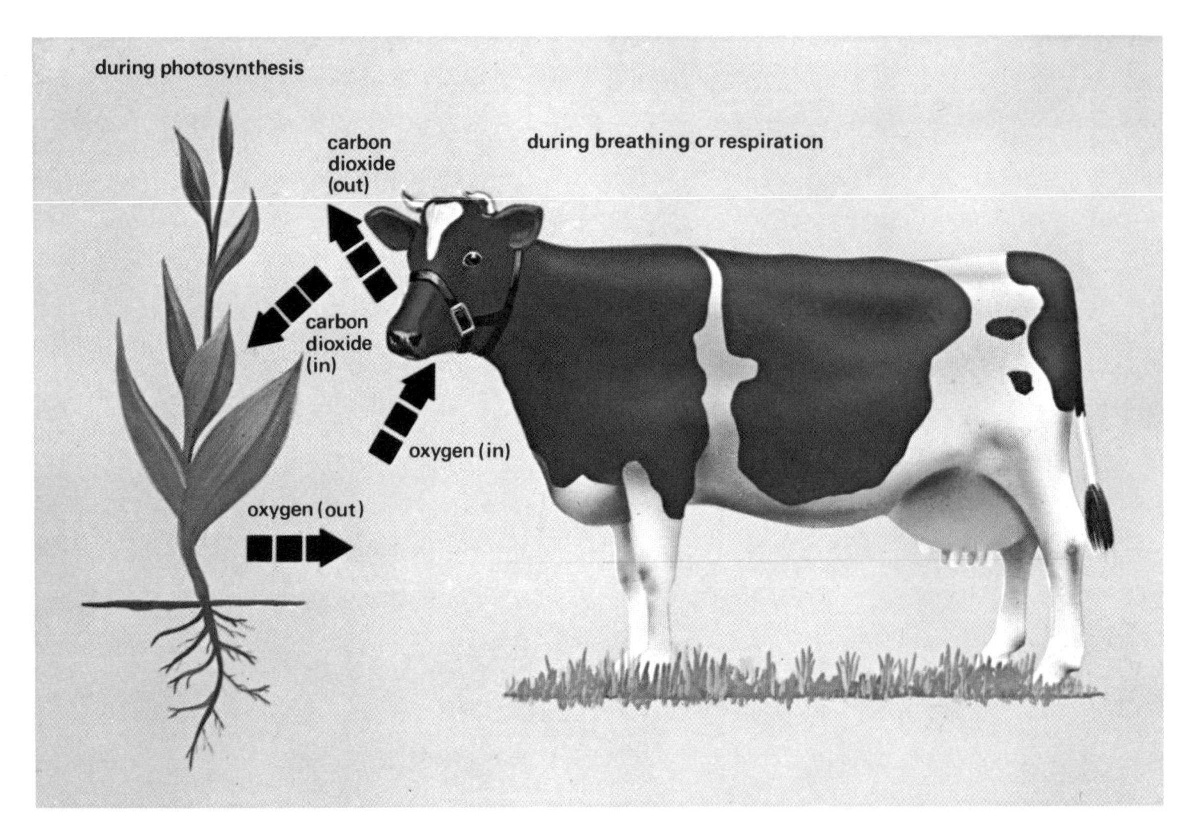

Photosynthesis is a very complicated process performed by green plants. Humans and animals breathe in oxygen and breathe out carbon dioxide. Plants take in the carbon dioxide and use it together with sunlight to make the food they need. A by-product of their feeding method is oxygen, which is given off from the surface of the leaves.

called photosynthesis. The plants use carbon dioxide and sunlight to help them make the food they need. They 'breathe out' oxygen. Without green plants on earth we could not live. Ozone, the gas that protects us from ultra-violet light, is composed of three oxygen atoms.

Other important substances found in air include moisture in the form of *water vapour,* and tiny specks of dust and salt carried into the air by wind blowing over the oceans.

Scientists have found that the atmosphere around the earth is different at different heights above the surface. These layers lie piled on top of each other until you reach space itself where there is no air. The layers of the atmosphere are known as the *troposphere;* the *stratosphere;* the *ionosphere;* and the *exosphere.*

The troposphere

Each layer of air that envelops the earth (right) has a different temperature and function.

From ground level, the troposphere extends upwards for several kilometres. In normal conditions, it gets cooler the higher you go, about 7°C for every kilometre you travel

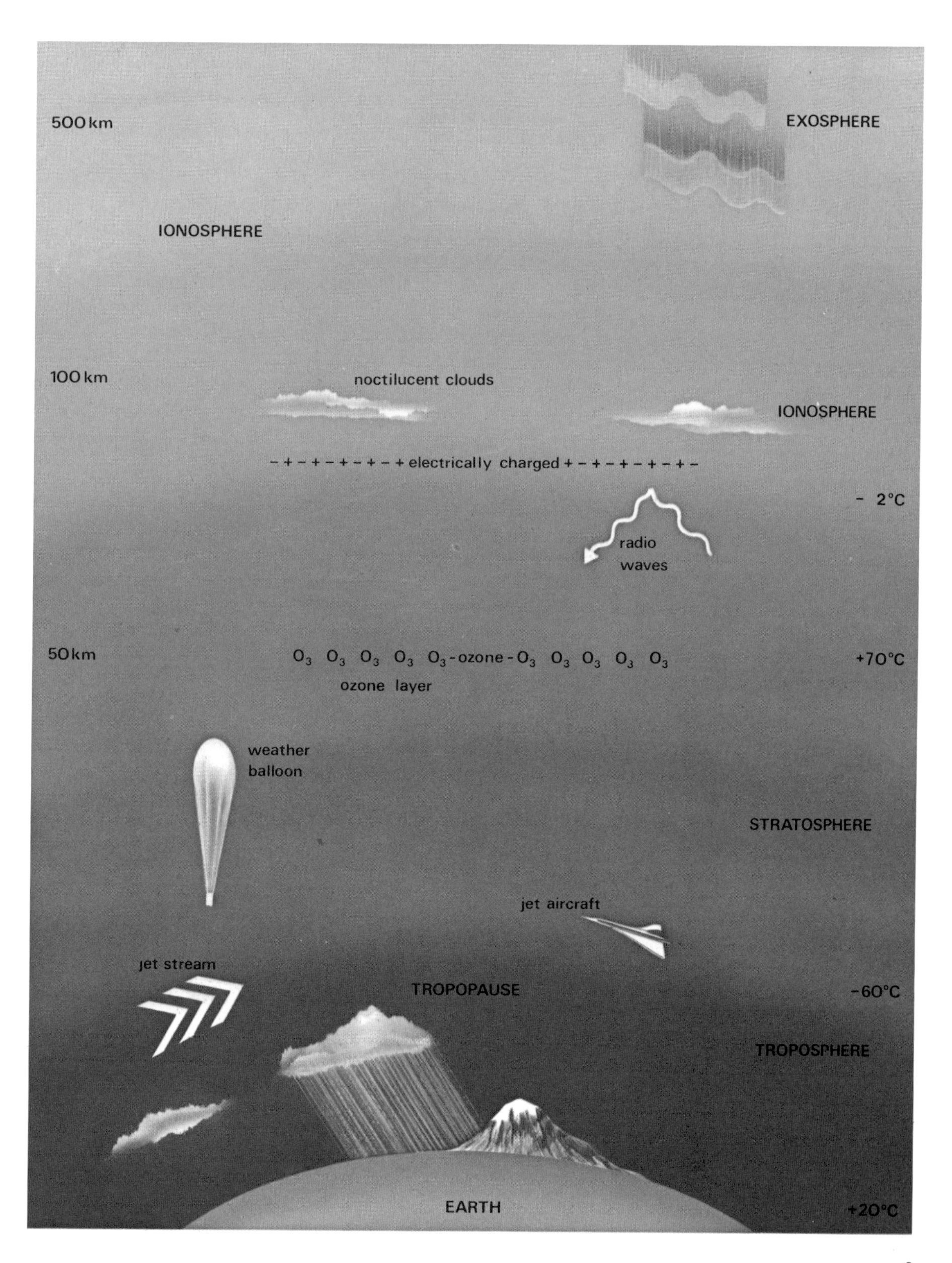
500 km
EXOSPHERE
IONOSPHERE
100 km
noctilucent clouds
IONOSPHERE
-+-+-+-+-+ electrically charged +-+-+-+-+-
- 2°C
radio
waves
50 km
O_3 O_3 O_3 O_3 O_3 - ozone - O_3 O_3 O_3 O_3 O_3
ozone layer
+70°C
weather
balloon
STRATOSPHERE
jet aircraft
jet stream
TROPOPAUSE
-60°C
TROPOSPHERE
EARTH
+20°C

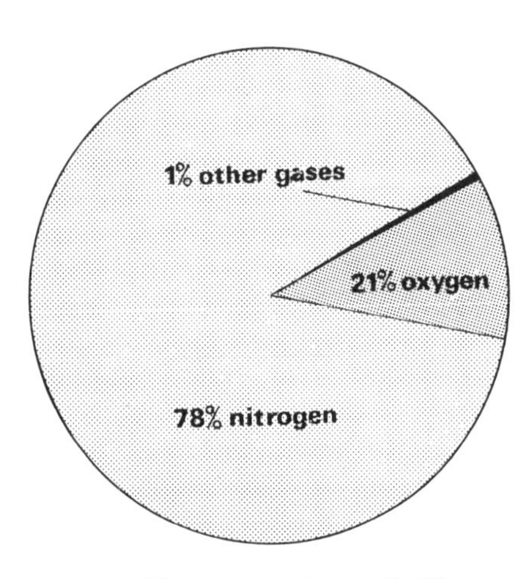

The proportions of different gases in the atmosphere are shown in this diagram. Oxygen, which is essential for life, is only quite a small part of the air we breathe.

upward. Finally the temperature stops dropping and becomes *stable,* that is, stays the same. This level in the atmosphere is called the *tropopause.*

The troposphere contains about three-quarters of the atmosphere. It becomes thinner and less dense the higher you go, that is, it becomes *rarefied.* The troposphere also contains more than nine-tenths of the moisture and other particles such as dust that are found in the air, so most of our weather begins in the troposphere.

For example, nearly all clouds are formed in the troposphere, and the major wind systems which affect weather operate only in the lower part of the troposphere.

Some winds called *jet streams* have been discovered in the high parts of the troposphere, often extending into the next layer. These fast-flowing air currents can travel at more than 160 km/hr so it is important to know about them when navigating an aircraft.

The stratosphere

Starting above the tropopause and rising to a height of about 80 km is the *stratosphere.* Above the tropopause the temperature first remains steady, then rises to about 7°C, then falls again to about *minus* 70°C.

Scientists think the rise in temperature within the stratosphere is probably caused by a layer or concentration of *ozone.* This gas acts as a filter, stopping some of the sun's rays called ultraviolet rays from reaching us. The tiny amounts that do reach earth give you a suntan, but too much causes painful sunburn. If the ozone layer did not block out most of the ultraviolet rays, they would kill all living creatures.

The stratosphere is a calm region by comparison with the troposphere, though it is sometimes disturbed by jet streams. Pilots often fly in the stratosphere to avoid disturbances in the troposphere. There are few clouds.

The ionosphere

This layer extends between about 80 and 500 km above the earth. It is called the *ionosphere* because the sun's rays have *ionised* most of the molecules or particles.

Industrial waste pollutes our atmosphere with unwanted and sometimes poisonous gases that wither trees before their time and blacken people's lungs. Fortunately we are now aware of the danger and hopefully the air will gradually clear.

A molecule is a tiny particle of a compound, and when it is *ionised* it is separated into the atoms of the elements that compose it, which become charged with electricity. Ionised air conducts electricity better than non-ionised air, so this layer is important for radio communication. Low frequency radio waves are reflected by this layer.

The exosphere

This is the outer layer of Earth's atmosphere and it gradually gets thinner and thinner until it gives way to outer space.

Air pollution

Smoke from fires and chimneys is one of the oldest forms of *air pollution*, that is, any impurities that are present in the air. The larger smoke particles eventually fall to the ground, but very small solid and liquid particles remain in the air and sometimes the smoke-filled air is trapped close to the ground. This causes the brown haze we call smog. If the air contains too much smog, you have difficulty breathing. This happened in 1952 in London, when the smog was so thick that it killed 4000 people. Air pollution also reduces the amount of sunshine reaching the earth's surface.

Today, many scientists and governments are working to control air pollution. In some places, people must use fuel which does not smoke. For example, when coal burns in an open fire it gives off a gas called *sulphur dioxide* which quickly corrodes or breaks down metal or stone as well as making the air less suitable for breathing. To avoid more disasters like the London smog of 1952 the burning of open coal fires has been restricted. Cars are being designed so they do not discharge harmful gases into the air. Many countries now have laws to control air pollution.

For many years, we have been using products in spray cans, such as hair spray, polish, perfumes and paint. Now scientists are finding that some of the gases which are used in these 'pressure packs' may be harmful to the ozone layer in the stratosphere. Since this layer is what protects us against the sun's ultraviolet rays, manufacturers are looking for substances to use in pressure packs which will not damage the ozone layer.

Weather

Weather is the day-to-day state of the atmosphere at any particular place. It is different from *climate,* which is a way of describing the usual weather of one place over a long period. The scientific study of weather is called *meteorology* and the people who do this work are known as *meteorologists.* Things that affect weather include air temperature and movement, air pressure, and the amount of moisture in the air.

Temperature

A dramatic view from Apollo 9 of storm-bringing cumulonimbus clouds forming over the Amazon Basin.

The atmosphere is always in motion. The energy which makes the air move comes from the sun. About three-fifths of solar radiation (the sun's rays) coming towards

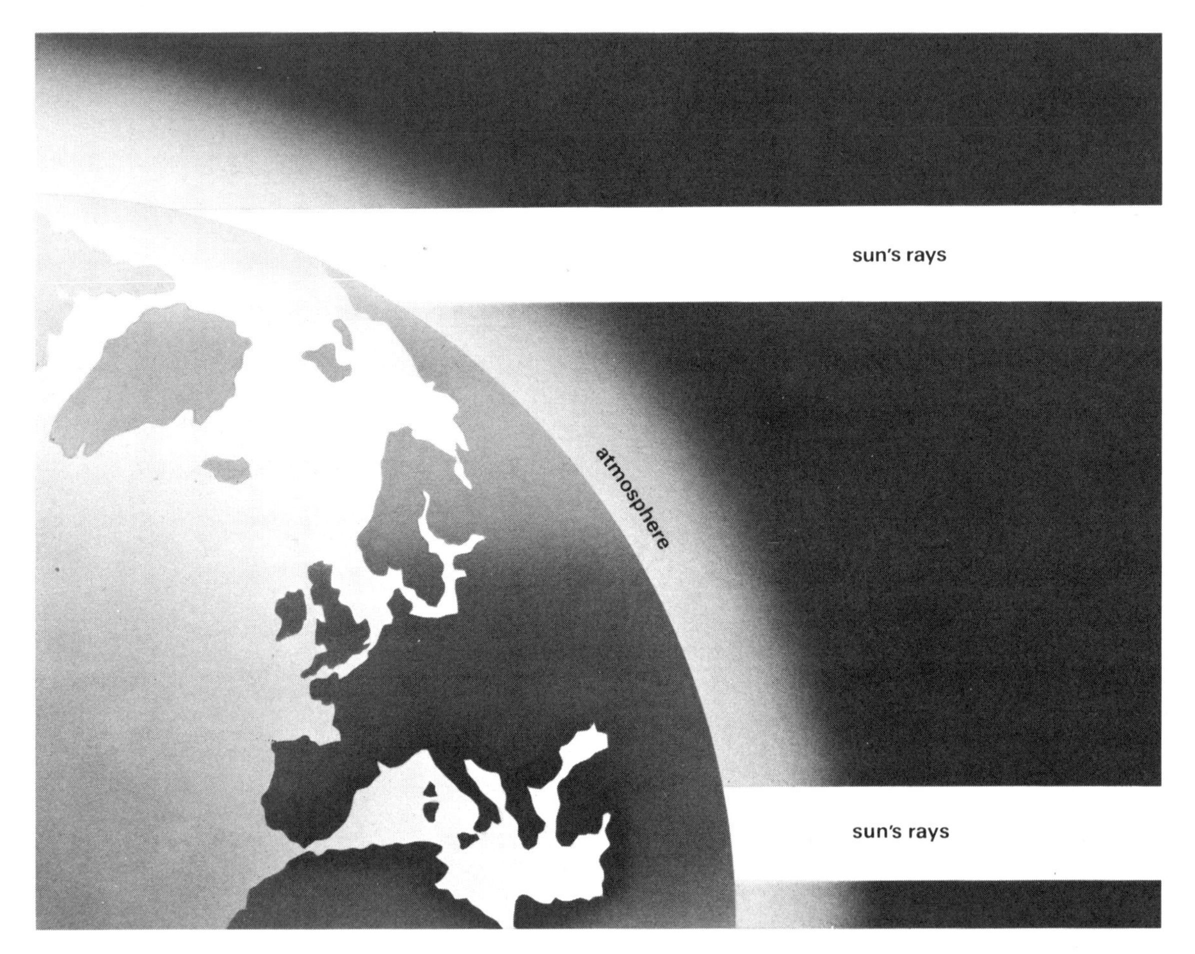

The sun's rays strike the earth directly and constantly only at the equator. The further north or south you go, the colder it gets as the rays strike at an increasingly oblique angle. Because the earth tilts on its axis during the year-long orbit round the sun, for some months the rays miss one or other of the Poles altogether and the long Polar night sets in. At the height of the Polar summer, however, the sun's rays shine on the region day and night.

the earth is absorbed by the atmosphere, so only a small amount of the sun's heat actually reaches us. The amount of heat reaching earth varies from place to place, which is why some countries are hot and others are much cooler.

The land around the middle of the earth – the Equatorial zone – receives the greatest amount of heat. The tilt of the earth on its axis means that, from the Equator to the Poles, the sun's rays must pass through an increasing thickness of troposphere before reaching the surface. They also strike it less directly. As you move further north or south of the Equator, the sun's rays bring less and less heat to the earth's surface.

Temperature is measured with *thermometers.* These instruments contain a tube of liquid, usually mercury or alcohol. When it gets hot, the liquids expand, or get larger, and rise up the tube. The tube is marked with degrees, so when the liquid stops, you can read how hot it is from the tube. When it is cold, the liquid contracts, or gets smaller, and drops down the tube to a lower level.

Pressure

Although you cannot see air, it has weight. If you weigh an open, empty container you are actually weighing both the container and the air inside it. We call the amount of air or anything else that is weighed, its *mass.* We say mass rather than weight because air expands when warm and it then has less weight than cold air. This is why warm air rises. The total mass of all the air around the earth is very great, about 5000 million million tonnes. You cannot feel the weight of all this air because it not only presses evenly on your body on all sides, it is also taken into the body through the lungs and so presses on you from inside. The air also presses on the earth's surface. This pressure is called atmospheric pressure.

In laboratories a pump called a *vacuum pump* may be used to pump all the air out of a container to show the existence of air pressure. If this is done with a thin container, like a soft-drink can, the pressure of the atmosphere crushes the can which has no pressure inside it. If the container is very strong a vacuum will be formed inside. A bell jar of thick glass or a metal sphere is usually used for this experiment.

Atmospheric pressure is measured by a *barometer.* This may be a large glass tube, closed at one end and standing in a bowl of liquid mercury. When the tube is filled with mercury, the open end is carefully placed below the level of the mercury in the bowl, and the liquid in the tube falls to a certain distance above the level of the liquid in the bowl. With changes in the pressure of the atmosphere, the column rises and falls in the tube. At sea level, the height of the column will usually vary between 736 and 775 mm according to the weather. Usually, low

A mercury barometer is more precise but more awkward to set up. A glass tube, marked off in millimetres, is filled with liquid mercury, inverted into a trough filled with the same fluid and clamped in place. The level of mercury should settle at about 760 mm, the average atmospheric pressure at sea level, but changes in the weather will raise or lower the level 30 or 40 millimetres.

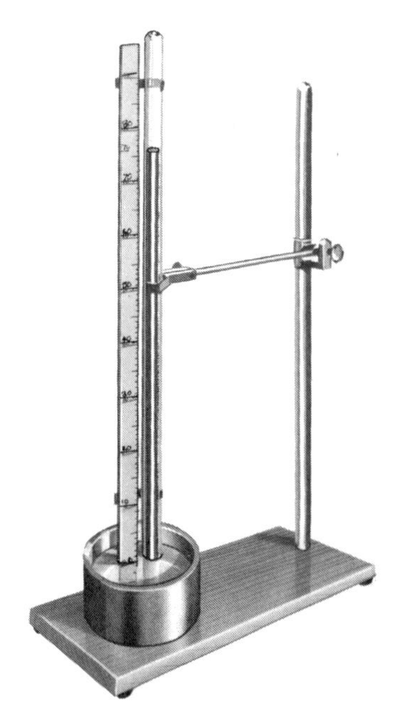

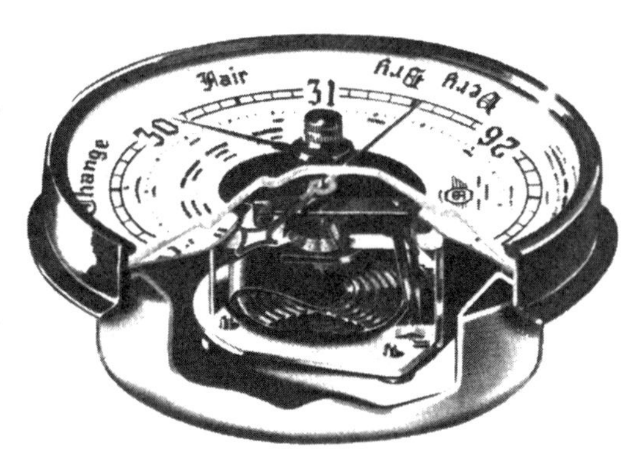

An aneroid barometer contains a metal box from which some air has been pumped. The box responds to changes in air pressure by expanding or contracting.

pressure is a sign of an approaching storm, and high pressure is a sign of clear weather. When a cyclone is approaching, air pressure usually drops quickly. Barometers are mostly used to give advance warning of storms and bad weather.

Today, the *aneroid* barometer is often used, because it is easier to take from place to place than a mercury barometer and is less easy to damage. The aneroid barometer contains an empty metal box with some of the air pumped out. Changes in air pressure make the box contract or expand, which causes a pointer to show pressure on a dial. This is the sort of barometer most people have in their homes.

Moisture

After hot day, the water vapour in the warm air condenses in the cool of the evening to form dew. It can form anywhere, but here the water droplets have condensed onto the framework of a spider's web, transforming it into a glittering basket of light.

Air always contains some water vapour. The amount of water vapour it can hold depends on the temperature. Warm air can hold more water vapour than cold air before it begins to *condense* or form drops of dew on a cool surface. Warm air containing a lot of moisture is called 'humid'. When warm air cools, it reaches a point at which water begins to condense out of it. This is called the *dew point,* and at this temperature the air is *saturated.* Of course, the dew point depends on just how much water vapour the air contains. It is measured by comparing the amount of water actually in the air at the time with the amount the air is capable of holding.

Moisture in the form of droplets in the air forms into clouds, dew and fog. When the clouds become too saturated to hold all the water, some of it falls to the earth as rain. Rain is collected and measured in rain gauges, which are containers graduated in millimetres. When the weather man tells us how much rain has fallen, he tells us in millimetres.

Weather forecasts

Weather forecasting – telling what the weather will be like for some time ahead – is very important to our daily lives, especially for ships and aircraft. If the weather forecast says there will be storms, sailors and pilots can get ready for them.

Weather forecasters prepare their reports from a mass of information collected and sent to them from weather stations. Meteorologists have instruments to measure conditions in the atmosphere and they also use radar and satellites to send back weather information from space. Most stations make four sets of observations each day, but some take readings of weather conditions every half-hour. The readings are then written in an international code and reports are sent by teleprinter or radio to weather forecasting centres.

At a forecasting centre, the information is decoded again and the details are plotted on weather maps. For example, the atmospheric pressures of all the weather stations at a particular time are plotted on maps. The forecaster or meteorologist then draws lines on the map to join all the places that have the same pressure. These lines are called *isobars.* Other information is given using symbols that are internationally accepted, to show areas of clear sky, heavy cloud, rain and so on.

Finally, such features as cold and warm *fronts* are added. A front is the leading edge of an air mass. The leading edge of a mass of warm air is called a *warm front* and when the mass of air is cold, it is called a *cold front.* Fronts usually bring clouds and rain.

When all the features are noted on the map, the forecaster has a summary or *synopsis* of the weather at a particular time, so the map is called a *synoptic chart.* At the same time, information about conditions away from the ground, in the upper atmosphere, is recorded on other charts. From these charts, the forecasters can prepare two

By measuring the depth of rain water that falls on a known area – such as the surface of this rain gauge – the average rainfall over a whole area can be calculated.

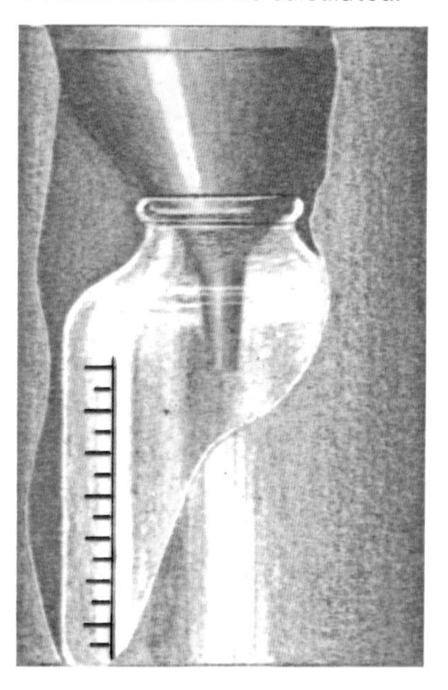

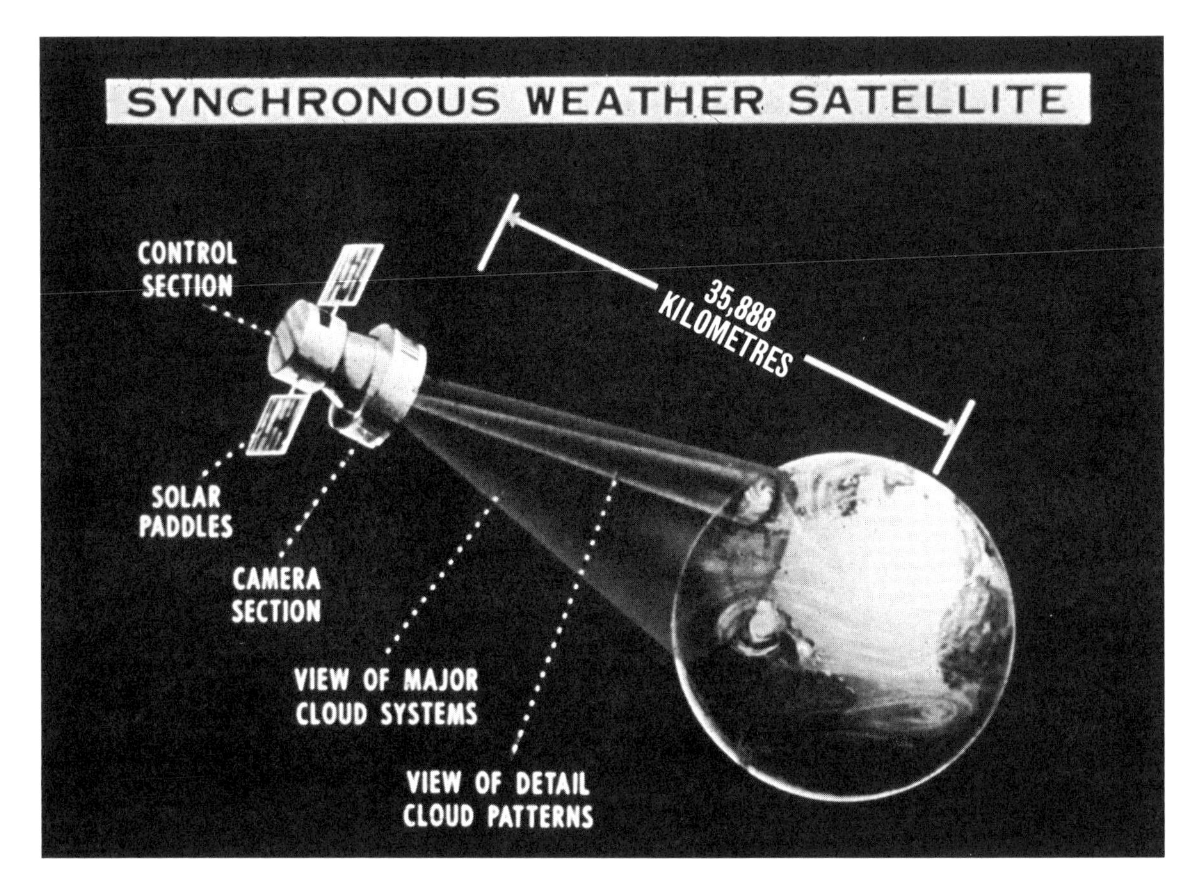

An orbiting weather satellite patrols the earth sending back information on cloud systems and patterns. Meteorologists interpret the information to give advance warning of changes in the weather.

kinds of weather forecast: short-range and extended forecasts.

Short-range forecasts tell you how the weather is likely to develop during the next 12 to 24 hours. Extended forecasts cover longer periods. A forecast which tells you about weather for the next five days is a medium-range forecast. There are also 30-day or long-range forecasts.

Clouds and winds

All air contains water vapour, lifted up by rising warm air from the seas, lakes and rivers. Warm air can hold more water vapour than cold air. Clouds are masses of tiny water droplets or tiny ice crystals in the atmosphere. When clouds form at ground level, we call them fog.

Because the temperature usually falls about 7°C every kilometre upwards from the earth, air that rises becomes cooler. As it gets cooler the water vapour condenses and turns back into droplets of water which collect around the specks of dust and other substances in the atmosphere. A mass of these droplets is a cloud. Clouds form different shapes in different weather conditions and scientists have given these shapes special names.

Types of clouds

Cumulus clouds get their name from the Latin word for 'heap'. These puffy white clouds heap up in thick masses, like mountains with flat bases and dome-shaped tops. They often form on hot summer afternoons about 1200 to 1500 m above the earth. If they become too heavy with

Once meteorological information has been gathered and decoded, the familiar weather map is produced. Lines called isobars are drawn through areas of equal pressure. Areas of high and low pressure can be seen on the map. Pressure is measured here in millibars.

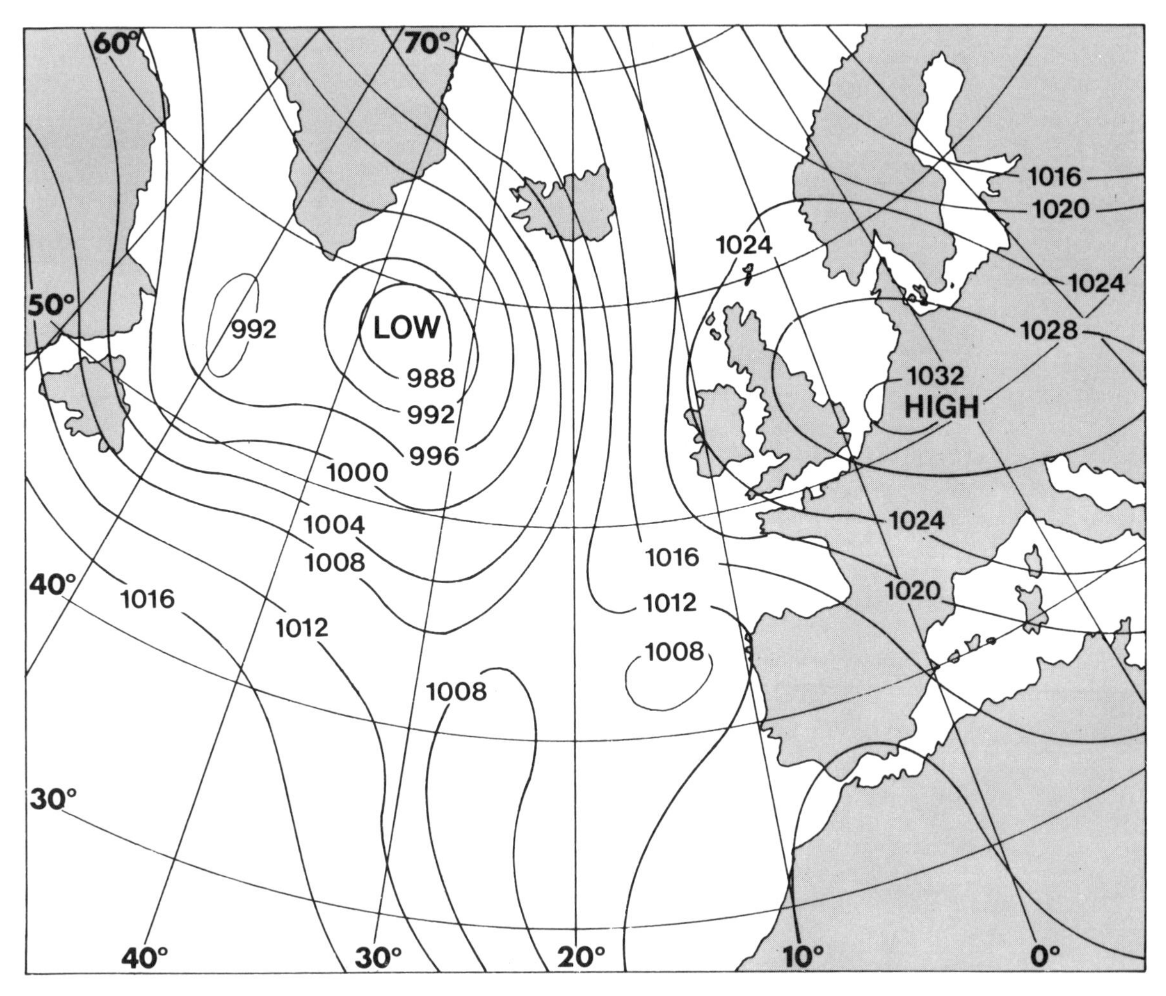

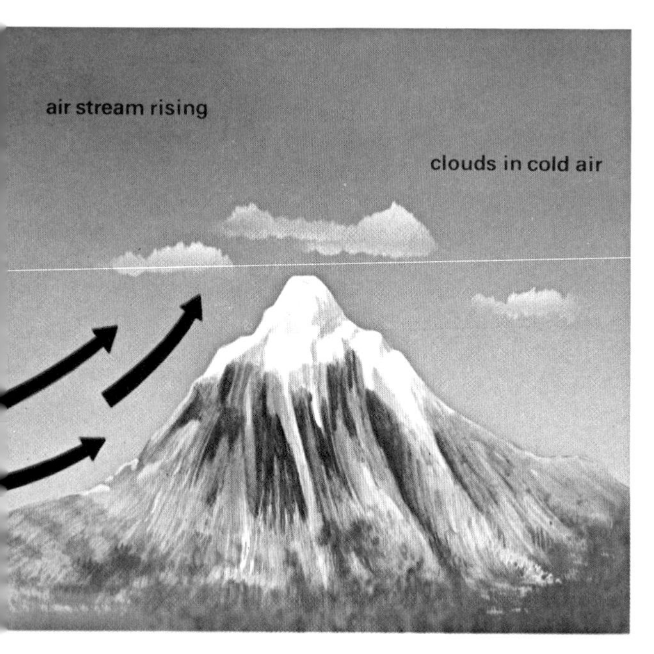

Clouds are made when warm air meets cold air, and the water vapour carried in the warm air condenses. Clouds form on mountain tops because warm air is funnelled up the mountain slopes to meet cold air at the top. When the sun shines on the sea, water vapour evaporates and rises with the warm air to form clouds when it reaches the cooler air.

water, they may turn into thunderclouds, so many of them in the sky can indicate rain.

Stratus clouds are wide, foglike clouds which appear low in the sky, from 600 to 1600 m. Their name comes from the Latin for 'spreading out' and they often mean bad weather ahead.

Cirrus clouds are the highest clouds in the sky and get their name from the Latin for 'curl'. They are little curly whisps of white ice crystals that are carried along quickly on the wind.

Nimbus clouds are rainclouds. Even their name means 'rainstorm'. They are dark, low overhead, and extend over much or all of the sky. Clouds may also form in combinations of these four kinds, such as *strato-cumulus* or *cumulo-nimbus.*

In very high clouds, when the temperature is very low, for example, minus 40°C, the water droplets turn directly into ice crystals and fall to earth as snow or hail.

There are four kinds of cloud (right), the round, puffy cumulus, low wide bands of stratus, nimbus the rainclouds and the high, wispy tendrils of cirrus.

Lightning and thunder

Lightning occurs when electricity travels between a cloud and the earth, or between different clouds, or between parts of the same cloud. Clouds become charged with the surrounding air. There is a difference in the amount of electricity in some clouds and other clouds or earth.

cumulus △

▽nimbus

stratus △

▽cirrus

Colour Library International

(Clouds are charged with negative electricity, earth with positive electricity.)

When the difference becomes great enough, the electricity leaps the gap between the charges in the clouds or between clouds and earth. The American scientist, Benjamin Franklin, was one of the first scientists to study lightning. He flew a kite with a metal key attached into thunderclouds. The electrified air made the key spark.

The lightning flash quickly heats the air it passes through, causing it to expand suddenly. This produces a *shock wave,* a rapidly moving wave of air which we hear as a thunderclap. As light travels very much more quickly than sound, we see the flash before we hear the thunder.

Lightning strikes when the negative charge in clouds builds up. The positive charge in the ground is attracted to it and rushes to meet it, usually channelling itself up a tree or a big building. 'Headers' of electrical charge reach out for each other, the charge building up all the time, until they meet and a huge spark is fired. Lightning conductors are metal rods, usually copper, which attract the charges to themselves to fire the spark safely. Trees, which have no such protection, are often blasted to extinction by lightning.

Winds

The movement of air around the earth is called wind. The fastest winds blow in *tornadoes* at speeds of up to 640 km/hr. Gentle winds are called *breezes.*

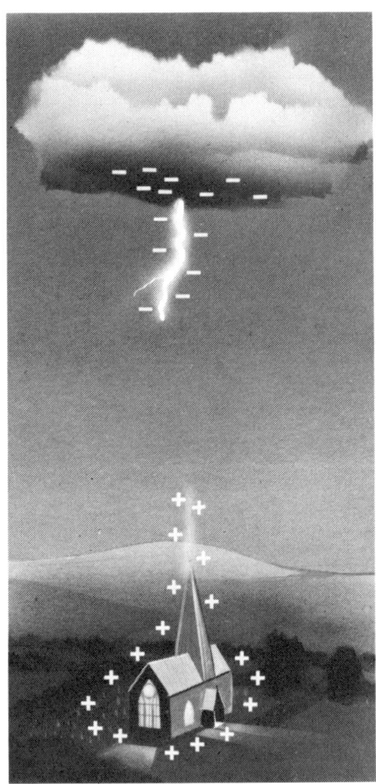

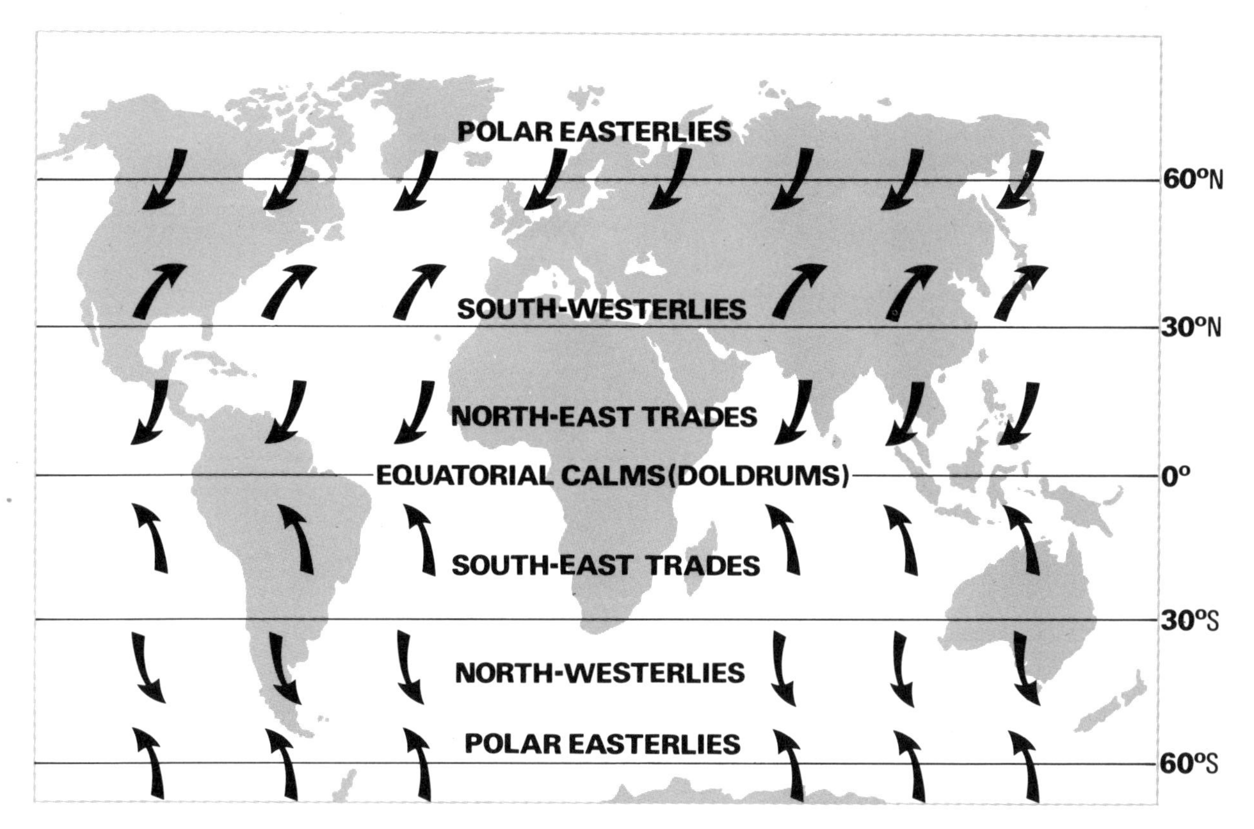

The latitudes and direction of the major wind belts of the world (above): It was essential to know these when world trade depended on the sailing ship, and they are still important to the meteorologist.

Sea breezes are created because the water heats more slowly than the land. When the warm air over the land starts to rise, the cooler air off the water moves inland to take its place. You can feel this movement, called an on-shore breeze, when you stand near a lake, the sea or the ocean.

The movement of the earth and the sun's heat causes a pattern of winds to move around the earth in bands known as the *chief wind belts.* These winds are called prevailing winds and they were very important to sailors in the days of sailing ships. The Trade Winds and the Roaring Forties (which blow around 40° south latitude) are examples. Though the general direction of these winds is constant, it will be affected by features such as mountain ranges, which create special local winds.

The Beaufort Scale

The strengths of winds are described according to a scale called the *Beaufort Scale.* This scale was invented by Sir

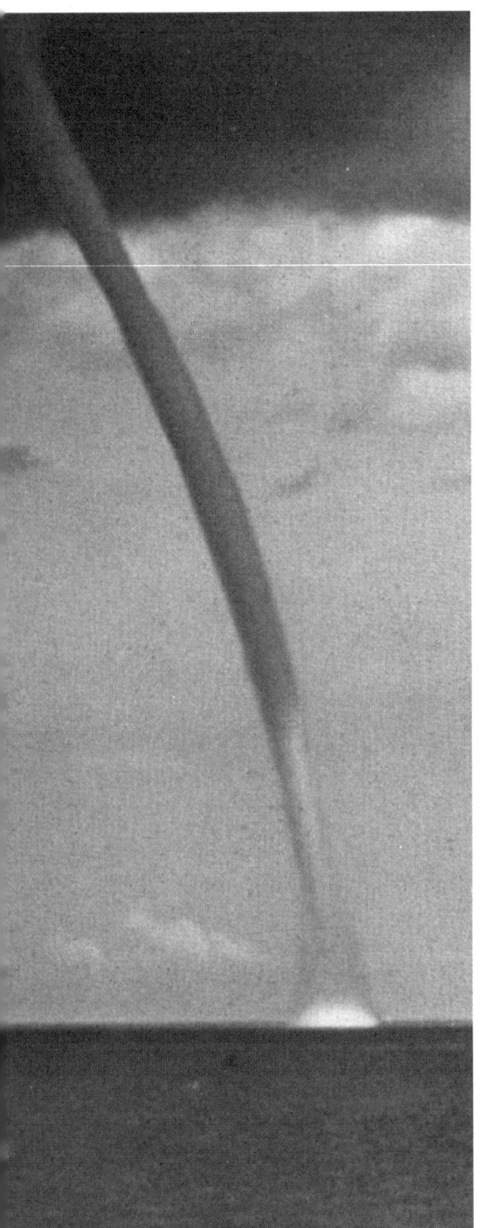

The whirling tube of a tornado, the kind of wind that tops the Beaufort Scale (above), devised by Sir Frank Beaufort as an international reference for wind strength.

THE BEAUFORT SCALE FOR WIND CLASSIFICATION

Beaufort Number	Speed mph	Description	Effects
0	0–1	Calm	Smoke rises vertically
1	1–3	Light air	Wind direction shown by drift of smoke
2	4–7	Slight breeze	Wind felt on face; leaves rustle; wind vanes moved
3	8–12	Gentle breeze	Leaves and twigs in constant motion; light flags extended
4	13–18	Moderate breeze	Dust and small branches moved; flags flap
5	19–24	Fresh breeze	Small trees sway; small waves on lakes and streams
6	25–31	Strong breeze	Large branches move; hard to use umbrellas
7	32–38	Moderate gale	Large trees sway
8	39–46	Fresh gale	Twigs break off trees; walking becomes difficult
9	47–54	Strong gale	Slight damage to houses (slates removed)
10	55–63	Whole gale	Trees uprooted; much damage to houses
11	64–75	Storm	Widespread damage
12	over 75	Hurricane	Violent conditions; sometimes loss of life

Frank Beaufort, who classified the different forces of winds by the way they acted over land and sea. His scale is based on the observation of how a gentle breeze makes trees move very slightly and makes only gentle foam appear on the sea, while at the other end of the scale a hurricane blows away everything in its path and whips the ocean waves into a fury.

High winds which we cannot feel on the ground include *jet streams* which blow high in the atmosphere, between the troposphere and the stratosphere, in the zone called the tropopause. Their speeds can vary from 160 to 500 km/hr. The positions and speeds of these streams are important because they indicate certain weather conditions to weather forecasters.

Satellites

Any body which orbits around another can be called a satellite, but the name is usually given to small bodies which orbit around the planets. The moon is a natural satellite of the earth. As well as the moon, artificial or man-made satellites now orbit the earth and some of the other planets in the solar system.

The first man-made object launched into space outside the earth's atmosphere was called Sputnik 1. It was launched in the Soviet Union on 4 October 1957. The second Sputnik was much larger. It weighed half a tonne and carried the first passenger, a dog named Laika. This satellite showed that a warm-blooded animal could live in

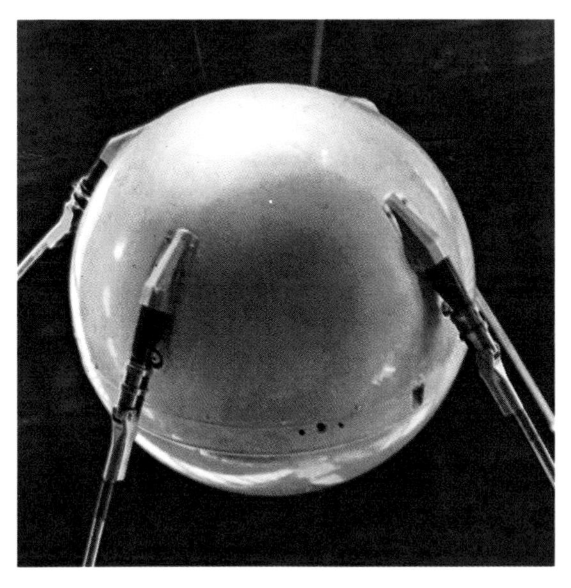

Barnaby's Picture Library

space. The satellite also sent back information about space and the upper atmosphere.

The first American satellite was launched in March 1958. Vanguard 1 was the size of a grapefruit and weighed only about 1.5 kg. By tracking its path, scientists discovered that the earth is very slightly pear-shaped. Since the space age began, about 1500 satellites have been launched. They are used for radio and TV communications, weather observation and defence. The first 'spy' satellites were launched in 1960. Their cameras take pictures of the earth's surface and they can detect rockets being launched.

Sputnik 1 (left), the first man-made satellite was launched by the USSR in October 1957. Sputnik is the Russian word for travelling companion. Telstar (right), the other satellite that gripped the public imagination was not launched until 10 July 1962.

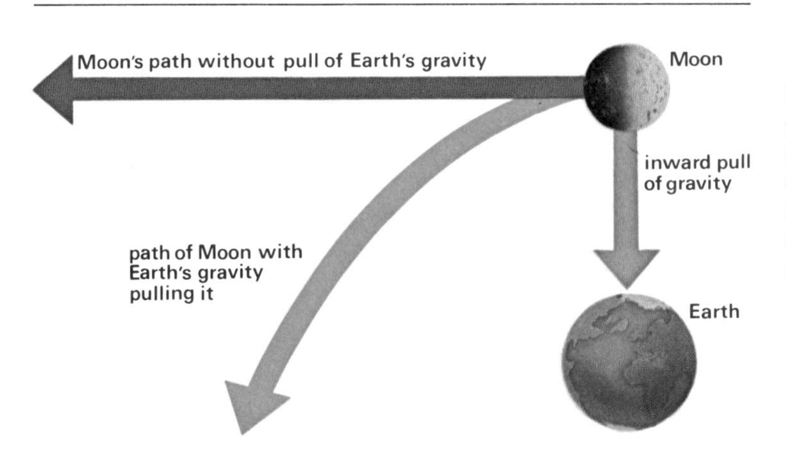

If the earth exerted a weaker gravitational pull we would be moonless. Gravity pulls the moon downwards as it tries to break free. To resolve these conflicting forces, the moon must orbit the earth.

Comsats

Satellites which are used to help people in different parts of the world to communicate with each other are called *comsats* which is short for *com*munications *sat*ellites. They can send live TV pictures to other nations, which means that you can see a programme from England in Australia at the same time as it is being performed.

Two kinds of comsats have been launched: *passive* comsats which simply reflect radio messages and *active* comsats which increase the strength of the message received and rebroadcast it. This is the standard type in use today.

One of the most famous communications satellites was Telstar, launched in July 1962. Telstar sent the first live TV pictures across the Atlantic. It orbited once around the earth every 158 minutes. When it was in view from two different stations on the ground, they could exchange visual messages.

As new and improved satellites came into operation, an international satellite corporation was formed by several governments. It was given the name 'Intelsat' and today Intelsat satellites send messages all over the world. One Intelsat satellite alone handles up to six thousand telephone calls at once between people in different countries.

Weather satellites

Satellites are used by meteorologists to help forecast the world's weather. They can also show forecasters where and when hurricanes will take place. Weather satellites now photograph the clouds over the earth.

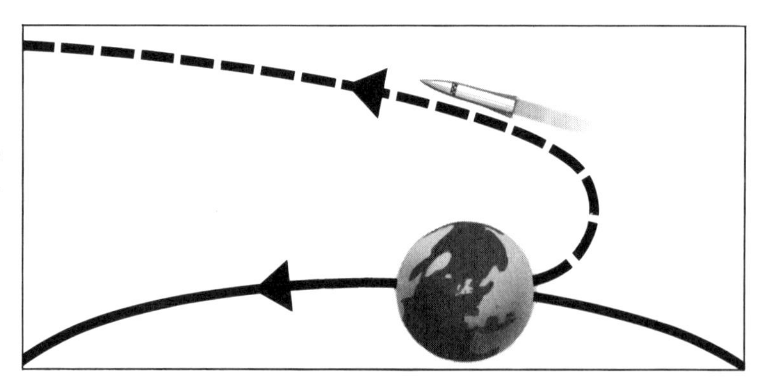

Rockets launched on earth must reach a speed called *escape velocity* to thrust them along fast enough to escape the clutches of earth's gravitational pull.

The southern hemisphere of the earth as seen from 35,000 km, showing cloud formation in the making.

Artificial stars

Ships and aircraft used to rely on stars in the sky to work out where they were. Today, there are artificial 'stars' called navigation satellites or 'navsats', which send out special signals. Because the satellite's position is accurately known, any vessel with the right equipment can pick up the signals and work out its true location.

Scientific satellites

Many satellites have been launched so scientists can study the physics of the earth, the space near the earth, radiation coming from space, and meteors. These observation satellites help scientists to understand how stars are formed.

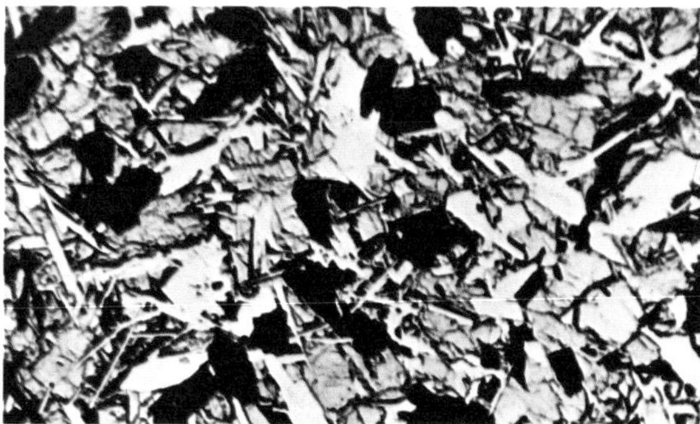

The footprints (above), left on the moon's dusty surface by an astronaut of the Apollo 12 moon flight, may remain there undisturbed for centuries. Samples of moon rock (above right) seen here through a microscope, were brought back by Apollo 11 astronauts for Earth scientists to study. The rock may hold clues to the formation of our planets.

By photography, satellites are also giving us a new look at the earth itself from outside. They can reveal such things as areas of crop diseases and good fishing grounds. They may show where to find valuable minerals and to locate areas of sea, land and atmosphere that are in danger of pollution. These satellites are called earth resources satellites.

The moon

In our solar system, there are 32 known natural satellites orbiting the planets. The moon is a satellite orbiting the earth; Mars has two small moons; Jupiter has 13; Saturn has ten moons and the largest of these, Titan, is the only natural satellite in our solar system which has an atmosphere of its own. We know that Uranus has five satellites and Neptune has two. The outer planets may also have small satellites that we do not know about yet. Occasionally, the gravitational pull of the planets may also 'capture' small asteroids as satellites.

Exploring the moon

The moon is the earth's nearest neighbour in space and men have always been fascinated by it. In the early 1600s, the great astronomer Galileo looked at the moon through his telescope and saw that there were vast dark plains, high mountains and great craters. More powerful modern telescopes showed the surface more clearly.

In the late 1950s and early 1960s, American and Russian space probes were sent up to explore the moon. The first Ranger probes sent back television pictures before crashing onto the moon's surface. Later probes

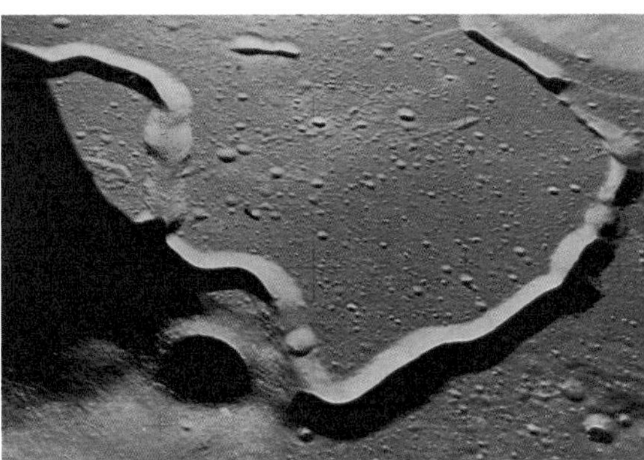

'softlanded' safely on the surface. Other spacecraft went into orbit around the moon taking photographs to make detailed maps of the surface.

In 1969, Apollo II was launched and the American astronaut Neil Armstrong became the first man on the moon when he stepped onto its surface on 20 July. He was followed by his fellow astronaut Edwin 'Buzz' Aldrin.

Later, several other two-man astronaut teams explored parts of the moon, carrying out experiments and collecting moon rocks to bring back to earth for study. In all, they

The Sea of Crises and the huge Langrenus crater (top) photographed from Apollo 8 as it approached the moon. Hadley Rille and the St George crater (below left) taken from the the command module of Apollo 15. St George crater (below right) seen from Hadley Rille.

collected 400 kg of moon rock, which revealed a great deal of new information about the moon.

Our moon is a large satellite compared with the moons which orbit the other planets. The moon's diameter of 3476 km is more than a quarter of the diameter of the earth. Because the moon is smaller than the earth its gravity is much less, about one-sixth as much. This means that an astronaut on the moon weighs only a sixth of his weight on earth. Because of this low gravity, the moon has no atmosphere, for it cannot keep any gases on or near its surface.

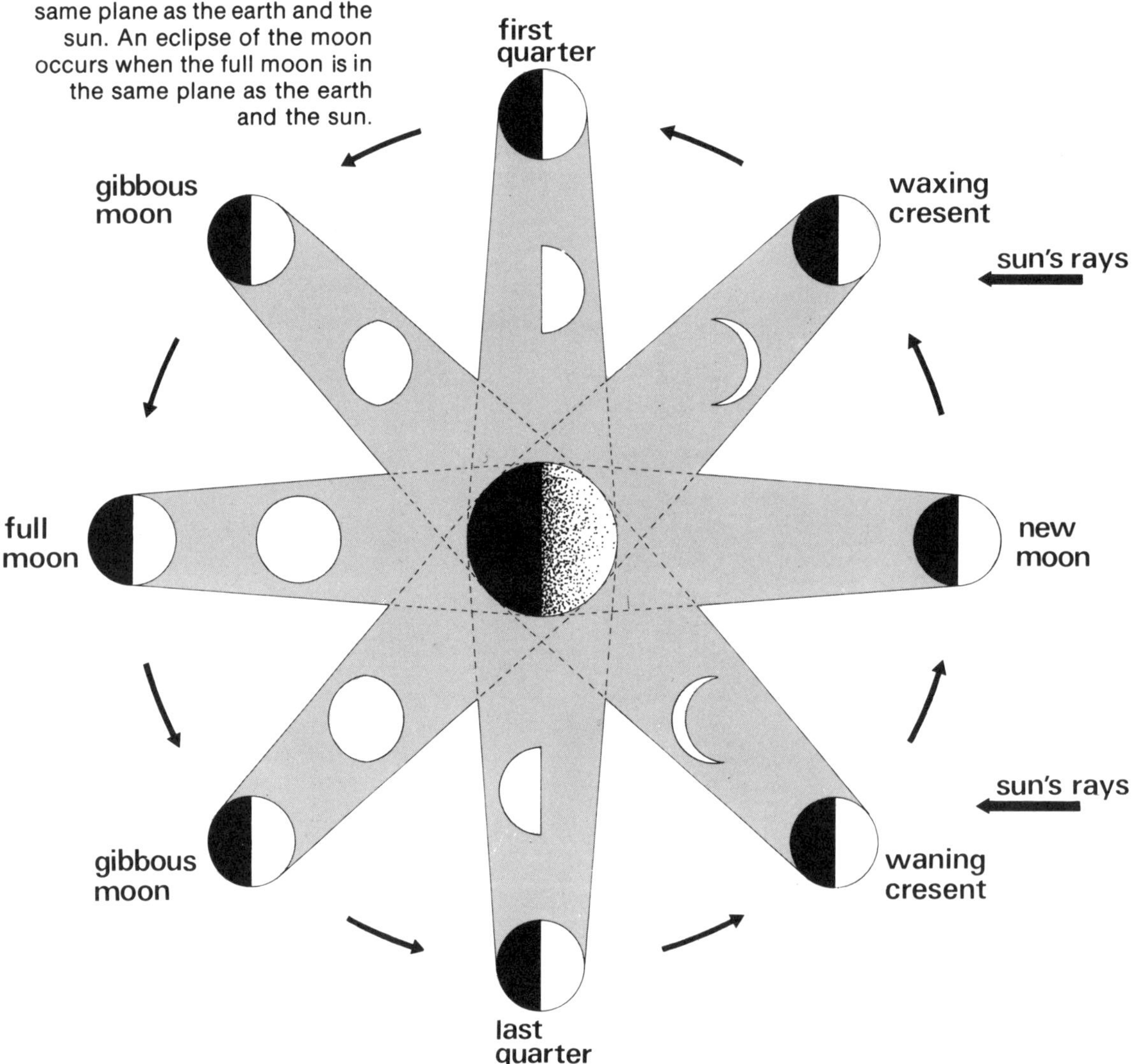

The changes in the apparent shape of the moon depend on how much of its illuminated surface is visible to us. The new moon is invisible because it is closer to the sun than the earth is and thus its illuminated side cannot be seen. As the moon moves in its orbit around the earth, we see the different phases of the moon.
An eclipse of the sun occurs when the new moon is in the same plane as the earth and the sun. An eclipse of the moon occurs when the full moon is in the same plane as the earth and the sun.

Without an atmosphere to insulate it, the moon has very high and low temperature extremes. During the moon's day (the *lunar* day), the temperature is over 100°C and at night the temperature falls below *minus* 150°C. On earth your shadow is fuzzy and grey but on the moon it would be sharp and clear because there is no atmosphere to diffuse or scatter the light. The moon has no changes of weather and no sound can be heard, for sound must travel through air or another medium. It cannot travel in a vacuum.

The moon travels around the earth about 384,400 km away. This orbit takes 27 days and eight hours. At the same time, the moon is turning on its own axis so we always see the same side, or face.

When a spacecraft travelled all the way around the moon, this was the first time anyone on earth had seen the other side, the dark side, of the moon.

The first manned landing on the moon was made by U.S. astronauts in Apollo 11 in 1969.

The moon's phases

Like the earth, the moon does not give out any light of its own. The parts of the moon we see lit up are reflecting light from the sun. This is why we see different amounts of the moon as it travels in its orbit; the changes in the shape of the moon as it is seen from the earth are called *phases* of the moon.

When the moon is in a position between the earth and the sun, the whole surface is dark to us, so we cannot see it at all. This is called the new moon. Soon, a slim curve of moon appears. It gradually grows, until about two weeks later the whole surface of the moon is lit up by the sun. This is called the full moon. Then the lit-up area gradually gets smaller and smaller until the next new moon is reached. When the bright area is increasing the moon is waxing, when it is decreasing it is waning.

A photograph of the crater Langenus from the Apollo 8 spacecraft. This crater is 85 miles in diameter.

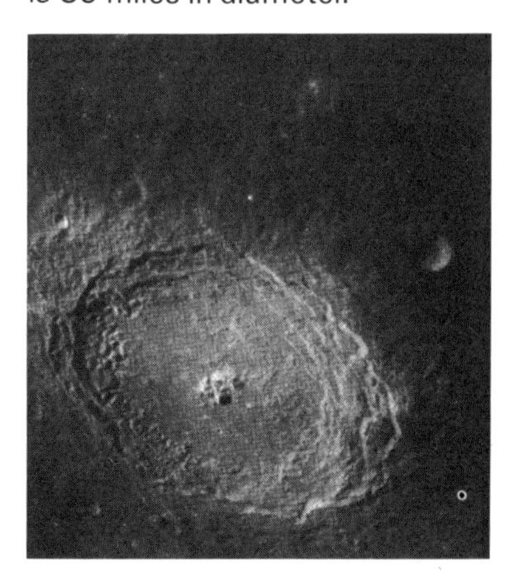

The moon's surface

When you look up at the moon, you can see dark and light patches. To small children, these patches form the face they call 'the man in the moon'.

The dark regions are called seas or *maria,* because early astronomers thought that they were full of water like

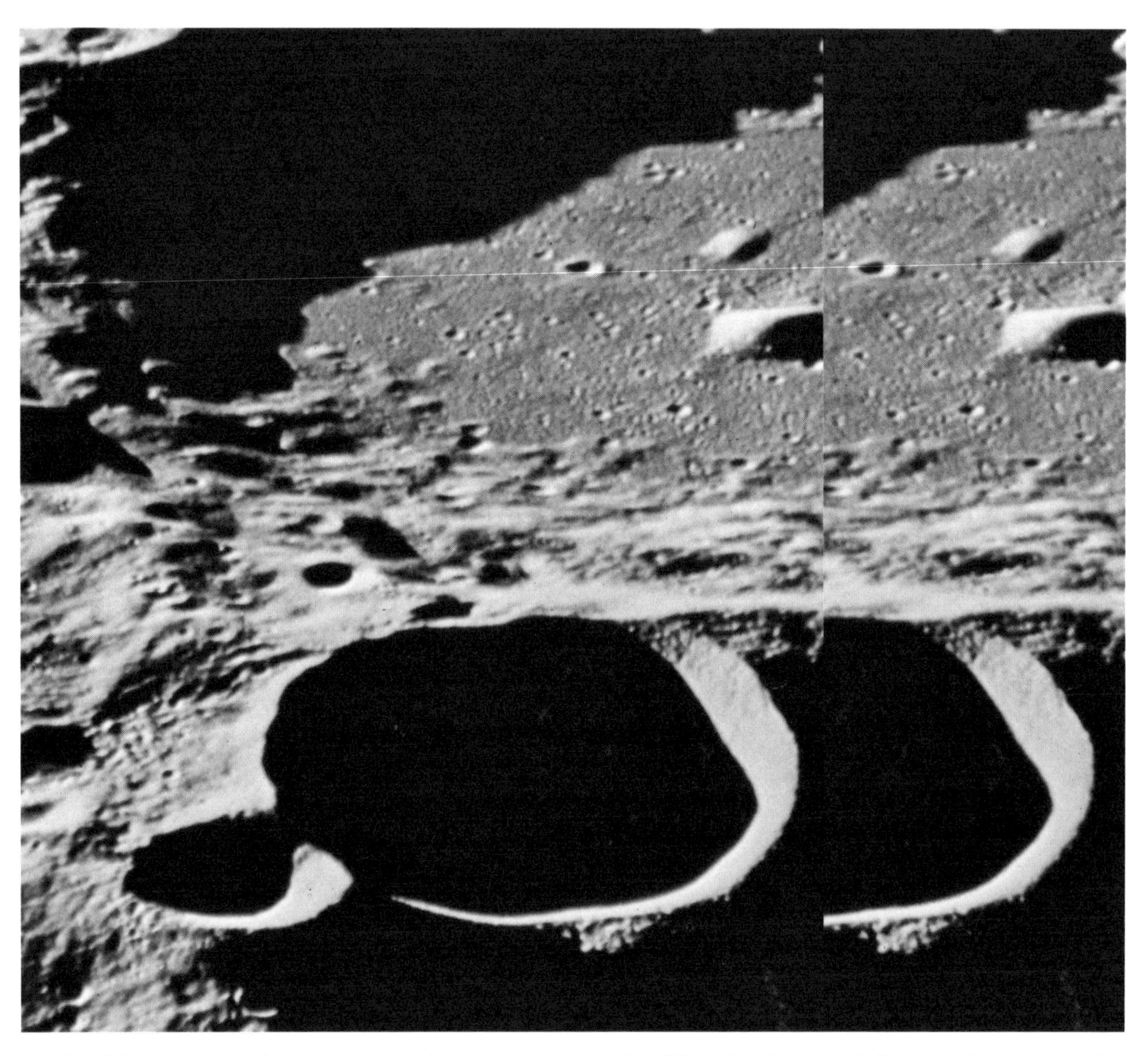

The big craters of the moon so resembled seas to the early astronomers peering through their telescopes that they named them all using the latin word for sea, *mare.*

the seas on earth. (The Latin word for sea is *mare;* the plural is *maria.* Latin was the international scientific language for early scientists). The light areas on the moon were named *terrae* or continents. The early names have been kept and are used by modern astronomers. The maria are low-lying, mainly flat regions. Some are nearly round and others are uneven or irregular.

The round maria mostly have a ring of mountains around them. Some examples are the Sea of Rains or Showers *(Mare Imbrium)* and the Sea of Crises *(Mare Criseum).* Some irregular maria include the Sea of Tranquillity *(Mare Tranquillitatis),* and the huge Ocean of Storms *(Oceanus Procellarum).* Most of these have no bounding mountain walls. Some of the terrae are very high, craggy mountain ranges.

Craters

These big, deep pits in the moon's surface are the main features of its landscape. The largest craters on the moon may reach 240 km across and 6000 m deep. Scientists think some were formed when giant meteors, fragments of rock in space, collided with the moon. The long chains of small craters on many parts of the moon's surface could have been caused by volcanoes on the moon.

Other interesting features are the *rilles* which are trenches hundreds of kilometres long. They cut through hills, craters and ridges. Scientists think some of these may have been caused by faults in the moon's crust. Some of the winding rilles seem to have been washed out of the moon's surface by rivers of lava.

The origin of the moon

By studying samples of rocks brought back from the moon, we now know it is older than anyone first thought. Most samples were between 3300 and 4400 million years old. Some rocks were made up of other rocks broken and stuck together by the force of meteors hitting the moon. The moon dust or loose surface soil is sprinkled with tiny fragments of glass where rocks have melted from the heat caused by meteors landing.

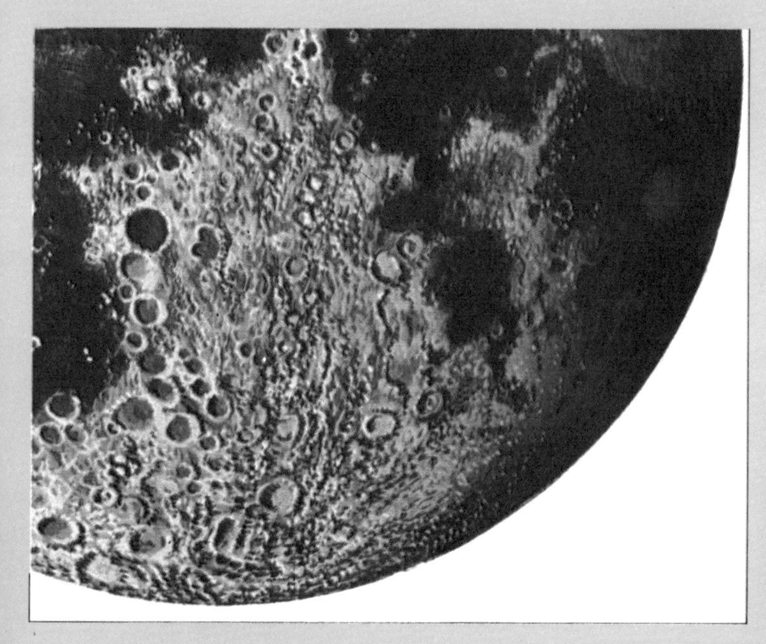

As the moon orbits the earth it also turns on its axis so that we only ever see one face of it. Until exploratory probes were sent up, the dark side of the moon remained a mystery.

Scientists can only guess what the inside of the moon is really like. Some say that it is cold hard rock. Others think it is hot and molten like the earth. It seems likely that the outer crust of the moon is a layer of jumbled rock between 95 and 240 km thick.

There are many ideas about how the moon was created. One belief is that it used to be part of the earth, and when it broke away, it left a large hole now filled with the Pacific Ocean. Other ideas are that the moon was formed at the same time as the earth by matter which gradually collected to become a solid body; or that the moon was a planet which was caught and held in place by the force of earth's gravity. Only much more study of the lunar rocks and visits to the moon itself will help us to find out more about our nearest neighbour in space.

FLIGHT

For centuries people dreamed of being able to fly but it was not until 1783 that the dream became possible, when the French Montgolfier brothers invented their hot air balloon, and a man named J. A. C. Charles designed a hydrogen balloon.

One of the earliest ways to get yourself up in the air was by hot air balloon. This mode of transport, which is still very popular, was pioneered by the French Montgolfier brothers, Joseph and Jaques, who gave their first public demonstration in June 1783.

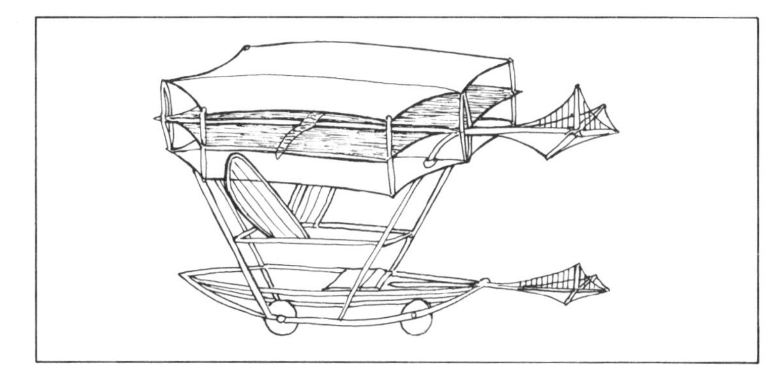

George Caley's design for a manned glider, built in 1849. It was rolled down a hill for launching and Caley's coachman was the reluctant test pilot.

Seventy years later, an English scientist, Sir George Cayley, found that objects heavier than air could also be made to fly. In 1804, he built a model glider and in 1849, he built a full-sized version in which his coachman became the unwilling passenger.

The Wright brothers

The next important development was the petrol engine which was light enough to power a flying machine. Two young Americans, Wilbur and Orville Wright, were already interested in flying and had practised with gliders. They built themselves a biplane called 'The Flyer' and fitted it

The very first manned flight in a heavier-than-air craft was made by bicycle manufacturers Orville and Wilbur Wright in December 1903 at Kitty Hawk, North Carolina. With Orville at the joystick, they made several flights, the longest of which was 245.6 m and lasted 59 seconds.

with a home-made gasoline engine. On 17 December 1903, at Kitty Hawk, North Carolina, they became the first people in the world to fly for any distance in a plane that was heavier than air and powered by an engine.

The Wright brothers flight led to the development of aeroplanes so rapidly that they were very widely used in World War I, especially after about 1916. But men were still interested in flying machines that were light enough to float in the air. Dirigibles, airships which could be steered, were becoming popular. In 1910, the German Count Ferdinand von Zeppelin introduced his giant metal-framed airship for commercial use.

In 1913, a Russian engineer named Igor Sikorsky built a huge four-engined aeroplane. Then, in 1919, the world's first commercial airline, British Aircraft Transport and Travel Limited, began operations from Croydon Airport just outside London.

Feats of flying

During these early years, many great and historic flights took place. By 1919, planes had flown over the Atlantic Ocean. In 1926, two pilots, Commander R. E. Byrd and Floyd Bennett, flew over the North Pole.

In 1927, Charles Lindbergh in his plane *The Spirit of St Louis* made his famous solo flight from New York to Paris, a distance of 5796 km which took him $33\frac{1}{2}$ hours.

Charles Lindbergh (centre) made the first solo flight across the Atlantic in May 1927, flying from New York to Paris in his plane *The Spirit of St Louis.*

The first air crossing of the Pacific Ocean was made in 1928 by the Australian flier Charles Kingsford-Smith. He flew his plane, the *Southern Cross*, from San Francisco to Brisbane in the total flying time of 83 hours.

Airships were still being used for transporting people and cargo by air. In 1929, the giant 236-metre airship, the *Graf Zeppelin*, flew 34,600 km around the world in 21 days eight hours.

Although airships were popular for many years, they were little used after the biggest one ever built, the 250-metre *Hindenburg* exploded while landing at New Jersey in May 1937.

Until the 1930s, aeroplanes were mostly made of wood and fabric. Gradually, they were replaced by metal-skinned planes. Some of the earliest designers of metal planes were Anthony Fokker of Holland and Junkers and Dornier of Germany. One of the first all-metal monoplanes was the Boeing 247, which carried ten passengers. It also had an automatic pilot and landing gear which folded, or *retracted,* into the body of the plane during flight.

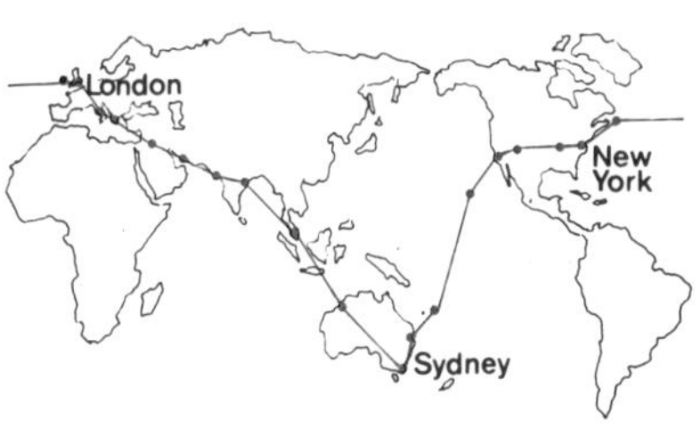

Charles Kingsford-Smith flew his *Southern Cross* from San Francisco to Brisbane.

The war in the air in World War I was fought with early fighters such as the Sopwith Camel for Britain and the Fokker, designed by a Dutchman but used by Germany.

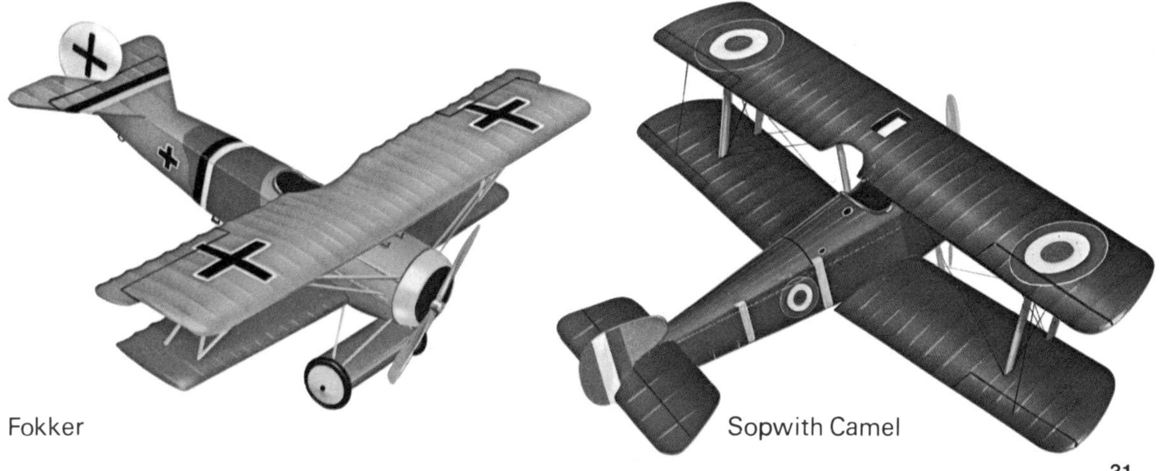

Fokker

Sopwith Camel

In 1935, possibly the most famous airliner of all made its first flight. This was the Douglas DC3, and even 30 years later there were still more of these planes flying than any other kind.

War in the air

Although planes were used during World War I, it was during World War II that aircraft became really important as weapons. Some of the most famous fighter planes were the Spitfire and the Hurricane which helped British pilots to win the Battle of Britain against the German Luftwaffe in 1940.

Jet engines had been invented before World War II broke out, but jet planes were not in general use until some years afterwards. Most of the battles were fought using aircraft with piston engines such as the Messerschmitt 109, dive-bombers such as the Junkers 87 Stuka, light bombers such as the De Havilland Mosquito and heavy bombers such as the Avro Lancaster, the Handley Page Halifax, the American B24 Liberator, the B17 Flying Fortress and B28 Super Fortress.

In World War II, air warfare became more sophisticated. More fighters were developed, including the Supermarine Spitfire for Britain, and Germany's Messerschmidt Me 109.

Spitfire

Paul Popper Ltd

Messerschmidt

Barnaby's Picture Library

The jet age

The jet age really started when in 1947 an American, Captain C. Yeager of the US Air Force, broke the sound barrier by flying faster than the speed of sound in a specially designed aircraft powered by a jet engine. Soon several countries were producing fast, powerful, jet-propelled aircraft.

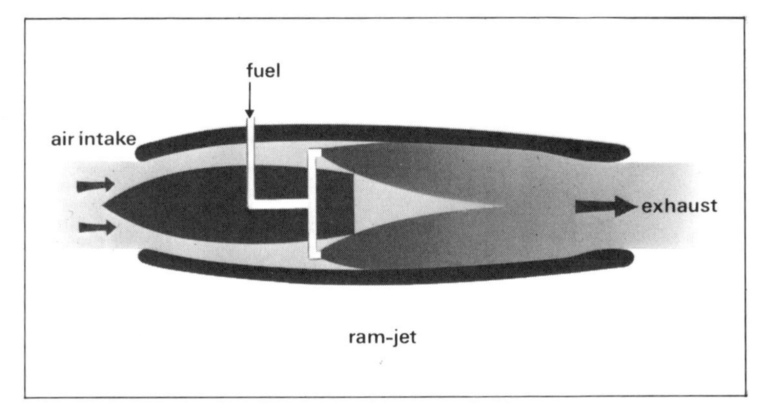

The ram jet is the simplest kind of jet engine, taking in air at the front and passing it to a combustion chamber where the fuel is ignited. The resultant blast of hot gases pushes out through the exhaust at the back, generating a forward thrust.

Central Press Photos Ltd

The English Electric Canberra jet was the first jet-propelled bomber.

Once space stations are built, we will need economic transport to get there. Once-only rockets would be too expensive. The space shuttle is carried up to a convenient height on the back of a returnable mother ship, from where it is launched into space.

Today we have aircraft which can fly three times as fast as the speed of sound. Modern supersonic passenger jet aircraft include the Concorde, designed jointly by England and France and the Russian Tu-144. The huge jumbo jets such as the Boeing 747 are subsonic but can carry up to 500 passengers at a time.

Special forms of transport between the earth and the upper atmosphere have already been tested, such as a space shuttle which is carried to a convenient height on the back of a mother ship. The shuttle takes off from its mother vehicle, orbits the earth, then returns and lands like an aeroplane, to be refuelled and launched again. This cross between a spaceship and a plane could be the commercial airliner of tomorrow.

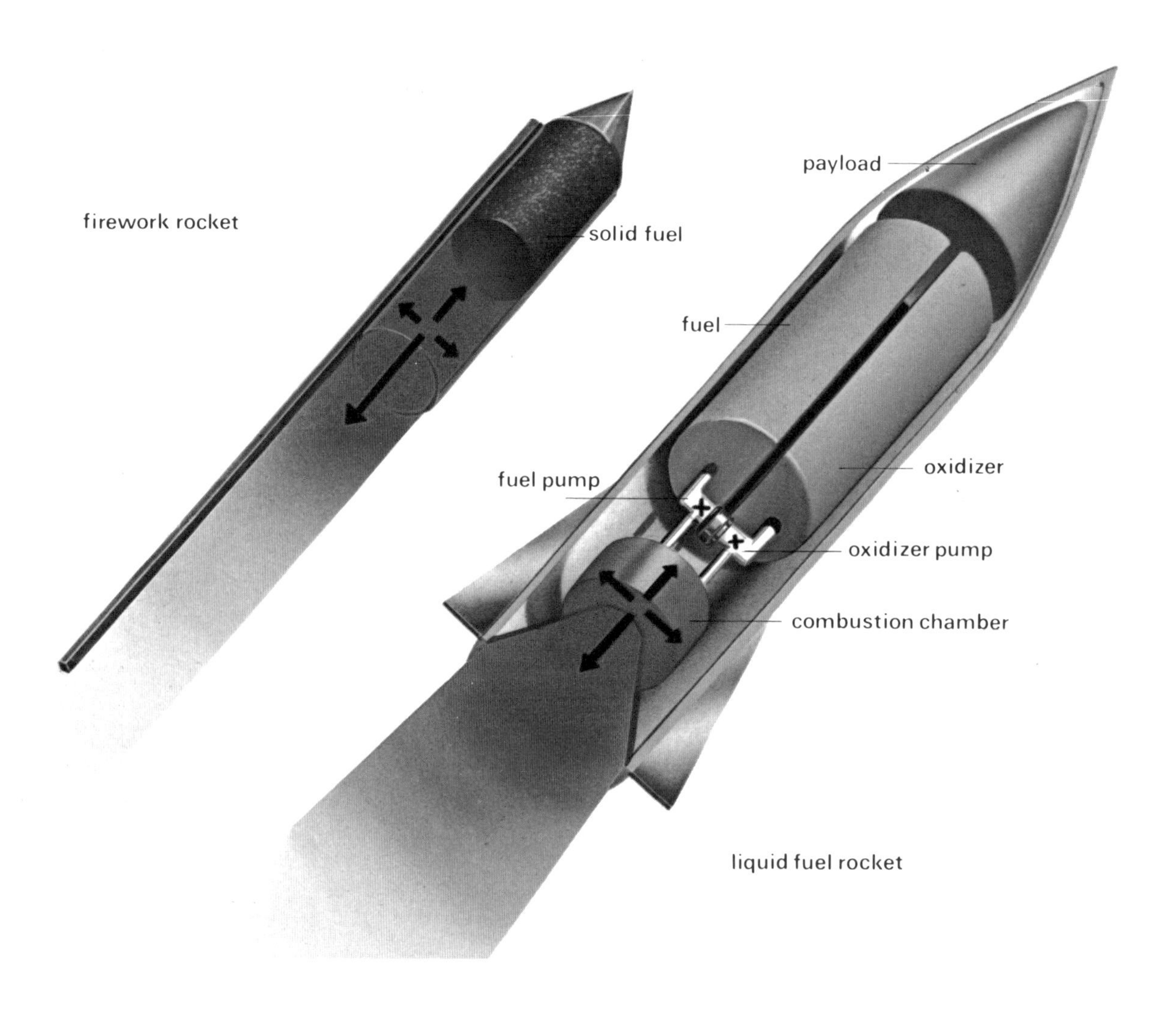

Rockets work by igniting fuel to burn at the base of the rocket to thrust forward. Simple rockets, like the firework kind, have solid fuel, but sophisticated rockets carry liquid fuel which means the thrust can be controlled by regulating the flow of fuel.

Rockets

Rockets have been known to man since the Chinese used them as weapons of war in AD 1252. Today they are still used as fireworks to celebrate special occasions. All rockets work on the same principle. Unlike aeroplanes driven by propellors, rockets do not need to be in air to function. They are therefore the natural choice for use in space.

In any rocket, fuel is burnt in a space called a *combustion chamber* which opens at the rear of the rocket. The fuel may be solid or liquid, but it must burn fast and produce a great deal of gas. This gas is usually passed out of a nozzle which is designed to give the gas stream the greatest possible speed. As the burning gas streams out of the nozzle at the rear of the rocket a *thrust* or push is developed at the opposite end which drives the rocket forwards. You can see this effect by blowing up a balloon, then releasing it in the air. As the air rushes out, the balloon travels forward. Space rockets work in a similar way, although they weigh several thousands of tonnes and need a great deal of fuel.

Many different fuels have been used for rockets. The most efficient is hydrogen and the rockets which fired the Apollo spacecraft on their way to the moon were powered by liquid hydrogen. Rockets used to launch space vehicles are usually run on kerosene. In both cases, since there is no air in space, the rocket must carry oxygen in solid or liquid form to burn or oxidise the fuel. Some rockets are driven by a fuel that has an oxidiser already mixed in it.

A solid fuel rocket is much simpler to design than a liquid rocket, but once started it cannot be stopped or slowed down. Rockets using liquid fuel are easier to control while they are running.

Huge rockets are needed to carry enough fuel to propel a capsule out of the earth's gravity field. This mighty multi-stage rocket is a Saturn V.

Barnaby's Picture Library

Multi-stage rockets

A great deal of power is needed to launch space vehicles and satellites. One rocket alone cannot carry enough fuel to provide all the power needed for the spacecraft to escape the earth's gravitational pull. Several rockets are used, one on top of the other to form a multi-stage rocket. As each stage is used up it drops off and the one above takes over, until only the spacecraft or capsule is left with small rockets to control its direction.

In 1949, the American space scientists launched their first two-stage rocket, which reached a height of over 400 km, setting a record for its time. By 1955, the United States had developed three-stage rockets powerful enough to launch artificial satellites. The Russians had also been working on rockets. On 4 October 1957, they launched the first artificial satellite. Over the next three years, both Americans and the Russians launched many satellites and space probes designed to send back information about the moon and space.

Paul Popper Ltd

Yuri Gagarin, who died in 1968 when he was only 34, was the first man in space. In *Vostok I* on 12 April 1961, he orbited earth once. Travelling at a height of 302 km and a rate of 27,840 km/hr, it took him 89.1 minutes.

The first men in space

The first man to travel into space was the Russian, Yuri Gagarin. He was called a cosmonaut (from *cosmos*, the Greek word for universe), while spacemen from the United States are called astronauts (from the Latin word for star, *astrum*). Yuri Gagarin travelled once around the earth in his spaceship, Vostok 1, on 12 April 1961. The first American to travel into space was Alan Shepard, who flew 185 km above the earth for a distance of 480 km.

By the end of 1963, four Americans and six Russians had travelled into space and returned safely. Among the Russians was the first and only woman (so far) in space, Valentina Tereshkova.

Problems of space travel

Before men could be sent into space, scientists had to solve the problem of bringing them safely back to earth. When re-entering the atmosphere, the spacecraft is moving so fast that friction with the air makes it extremely hot. So the capsule containing the astronaut must be protected by a thick heat shield. The capsule is also slowed down by parachutes which open during the last few thousand metres of travel towards the earth. At present, many spacecraft land in the sea (this is called a splashdown) for extra safety.

People who travel into space have to cope with problems not found on earth. One of them is *weightlessness.* An astronaut, except when his ship's rockets are firing, is always falling freely. This 'fall' may be upwards, downwards or sideways, but because his body has nothing to push against he is not aware of his own weight. When the rockets are fired, the ship's weight pushes against its own exhaust jets and part of the ship becomes 'down' to the astronaut who will once again be able to feel some of his own weight. But when the rockets are not firing, the astronaut is weightless, or in 'free fall', and will float and bob around inside the capsule. This can cause a kind of travel sickness if the astronaut has not been specially trained to expect it. This is one reason why astronauts have to be very fit and strong both physically and mentally. They go through a very long and tough training programme. While they are in space, instruments

The spacesuit (right) is the astronaut's whole world. It is tough enough to withstand small meteors, insulates him from extreme heat or cold maintains constant air pressure inside, provides oxygen and eliminates waste products.

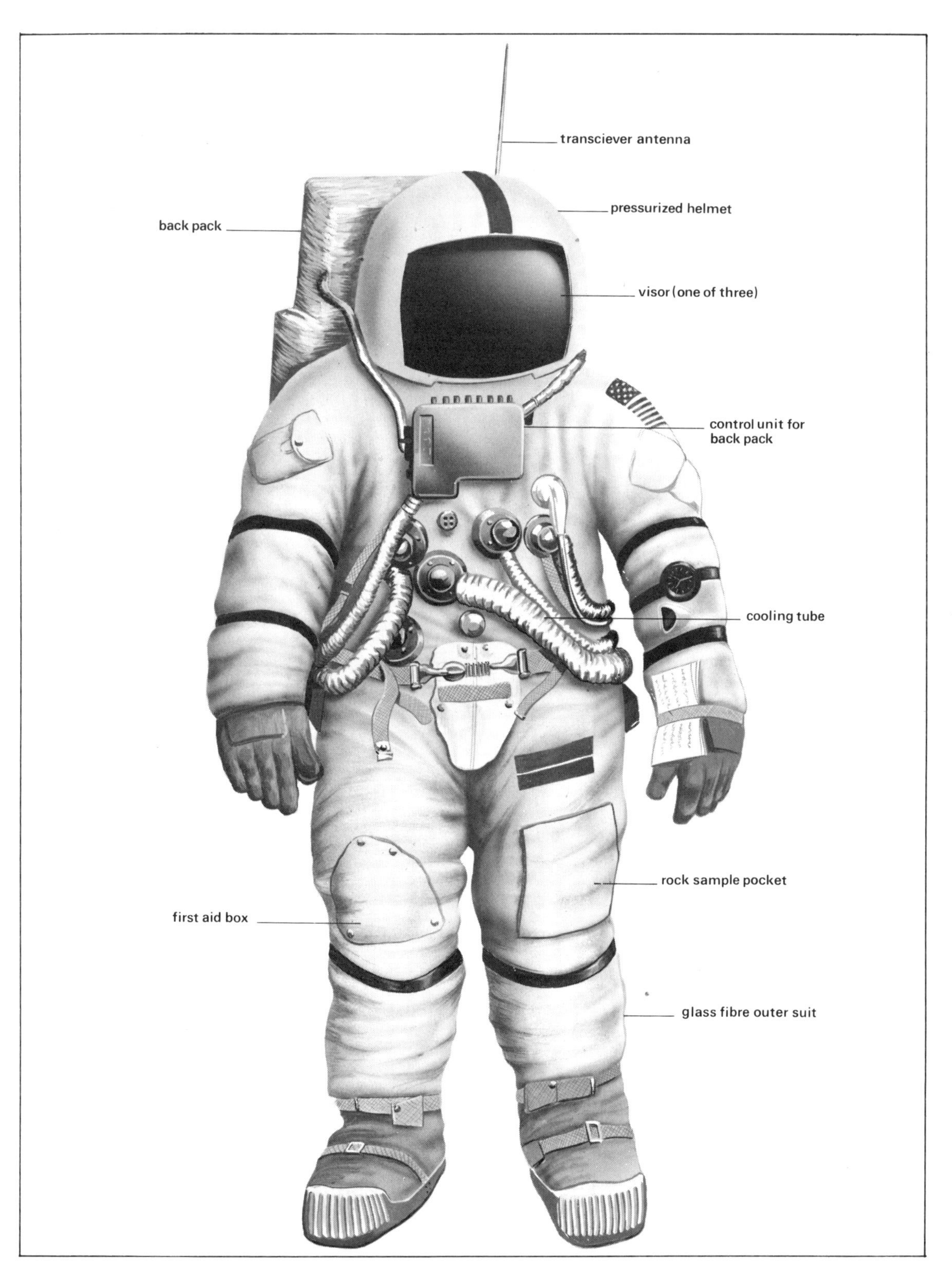
transciever antenna
pressurized helmet
back pack
visor (one of three)
control unit for
back pack
cooling tube
rock sample pocket
first aid box
glass fibre outer suit

Ed White space walking outside the capsule on the Gemini 4 mission of 1965. He manœuvred about with a hand-held jet and was connected to the capsule oxygen supply by a gold plated line.

or *sensors* attached to their skin send back information so doctors on earth can check on their condition.

Astronauts have to take with them everything they need to survive including food, water and air, so they wear spacesuits which provide these things. The suit not only has to be tough so that it doesn't tear. It must be insulated from the heat of the sun and provide the astronaut with a supply of air and a system to dispose of the body's wastes. The astronaut must also be able to move about while in it. The spacecraft in which astronauts travel also has a *life-support system* which does the same job as a spacesuit.

In weightless conditions, it is impossible to drink from a cup because the liquid floats in the air, so water is drunk from a tube like a straw. In early space flights, food was packed as a paste in tubes like toothpaste tubes. In Apollo flights, the astronaut had frozen foods to eat. By adding hot or cold water they could prepare a meal.

The Apollo projects

Project Apollo was started in 1961 by the USA. Its aim was to send a man safely to the moon and back. The project took nine years. New styles of rockets and many other new systems had to be built for Project Apollo.

The plan was to send a team of three astronauts in a three-part spacecraft. The spacecraft was made up of a command module for the crew, a service module to carry most of the equipment, and a lunar excursion module, or LEM, to land on the moon. The whole spacecraft was designed to orbit around the moon. Then two astronauts would climb into the LEM, which would separate from the main spacecraft and land on the moon's surface. Later it would return to orbit and link up with the main part of the craft. The LEM would then be left behind while the rest of the spacecraft came back to earth.

Scientists and astronauts still had problems to solve before the trip to the moon was possible. In January, 1967, during a practice countdown of the last few seconds before launching, a fault in the spacecraft caused a fire and the three astronauts, Virgil Grissom, Edward White and Roger Chaffee, were burned to death in their cabin. The spacecraft had to be redesigned to prevent other accidents like this.

In November 1968, testing started again and in July, 1969, astronaut Neil Armstrong took his first step onto the moon's surface. He was followed by Edwin Aldrin. The

Apollo 12 made the second moon landing, leaving relay equipment to stand forever rustless in the Sea of Storms.

A capsule takes off in grand style. Apollo 15 was launched in July 1971 by a giant Saturn V rocket. Splashdown (right) and retrieval from the sea are altogether less spectacular.

third member of the team, Michael Collins, stayed behind on the main spacecraft orbiting around the moon. The historic journey to the moon was successfully completed when the three astronauts returned to Earth safely in the command module and splashed down in the Pacific Ocean where they were picked up by the USS *Hornet.*

Other flights to the moon took place in 1970, 1971 and 1972. The manned flight of Apollo 17 which went to the moon in December 1972 may be the last which will take place this century. However, unmanned spacecraft or probes are still sent to the other planets. These probes include the Mariner flights which sent back first pictures of the surface of Mars. Man's adventures in space are only just beginning.

INDEX

THE EARTH'S ATMOSPHERE 1-28
FLIGHT 28-33
SPACE EXPLORATION 34-40

Page numbers in italics refer to a diagram on that page.
Bold type refers to a heading or sub-heading.